Gregor Sahler

China's crusade for African oil on the Example of Sudan

GRIN - Verlag für akademische Texte

Der GRIN Verlag mit Sitz in München hat sich seit der Gründung im Jahr 1998 auf die Veröffentlichung akademischer Texte spezialisiert.

Die Verlagswebseite www.grin.com ist für Studenten, Hochschullehrer und andere Akademiker die ideale Plattform, ihre Fachtexte, Studienarbeiten, Abschlussarbeiten oder Dissertationen einem breiten Publikum zu präsentieren.

Document Nr. V181809

Gregor Sahler

China's crusade for African oil on the Example of Sudan

GRIN Verlag

Bibliografische Information der Deutschen Nationalbibliothek: Die Deutsche Bibliothek
verzeichnet diese Publikation in der Deutschen Nationalbibliografie; detaillierte bibliografische Daten sind im Internet über http://dnb.d-nb.de/ abrufbar.

1. Auflage 2010
Copyright © 2010 GRIN Verlag
http://www.grin.com/
Druck und Bindung: Books on Demand GmbH, Norderstedt Germany
ISBN 978-3-656-05059-9

SS 2010

China's Crusade for African Oil on the Example of Sudan

European Energy Security

Gregor Sahler

Universität Hamburg

TABLE OF CONTENT

1. Introduction

Since the formation of the People's Republic of China in 1949 its political elites have cultivated diplomatic relations with many African countries. Throughout the decades, these diplomatic ties more and more entailed economic activities. When Deng Xiaoping took over the helm of the state at the end of the 1970s, China entered into a still lasting phase of unprecedented economic growth. To further fuel this upsurge, China soon had to start to import certain resources from abroad. By the mid-1990s Beijing finally realized that it would no longer be able to rely exclusively on its domestic oil reserves.

This paper illustrates China's oil-related activities on the African continent. To provide a clearer picture how the People's Republic proceeds in securing African oil, I decided to illustrate China's oil policy in Africa on one specific example. Therefore, I put Sudan in the center of my analysis. Sudan is one of China's most important oil providers; up to 80 percent of Sudan's daily oil produce goes to China.[1] Nevertheless, political instability, humanitarian crises, and revolting tribes account for a difficult investment environment.

To provide essential background knowledge about the topic, I will first present important facts and figures about China's and Africa's oil industry. I will, furthermore, give some perspective of how deeply China is actually invested in the African oil market.

In the third chapter, I will go into the details of China's and Sudan's oil diplomacy. Thus, it is important to put their relationship in an historical context. I will, thereafter, explain Beijing's approach in securing Sudan's oil reserves and I will shed some light on the question how the Chinese leadership justifies its engagement with a rogue state.

Before I draw a conclusion, I will outline some threats for China's strategy in Sudan. Next year the southern part of the country may secede from Khartoum. This scenario can be either threat or chance for China's crusade for African oil.

In this paper, I want to look into the questions, why China chose to go to Sudan considering that there are countries, even in Africa, with much larger proven oil reserves? Are there any drawbacks associated for the Chinese in cooperating with Khartoum? And did China's strategy in Sudan so far pay off?

---

[1] Engdahl, 2007

2. Facts & Figures about Sino-African Oil Relations

2.1.   The Institutional Framework of the Chinese Oil Industry

China's energy infrastructure has been and will be strongly dependent on oil. Today, the country relies to 40 percent on oil imports, but analysts suggest that this number will climb up to more than 60 percent in the years to come. Given the steady economic growth and the gradual depletion of its domestic reserves and considering the fact that a substitution of oil with more efficient fossil fuels,[2] nuclear energy, or even renewables seems to be out of reach in the short run, the People's Republic will have to tap new sources of oil abroad.[3]

After realizing, in the mid-1990s, that energy self-sufficiency has been an unrealistic dream, China decided to go abroad and compete in the international energy market. To successfully challenge the market power of the international oil companies that had a head start of 50 years in operating globally, China had to funnel its existing sinews. Therefore, China restructured the preexisting state oil and gas companies into two major enterprises: the China National Petroleum Corporation (CNPC) and the Chinese National Petrochemical Corporation (Sinopec). Both companies produce, import, trade, and process oil, albeit to a different extent. Together with the China National Offshore Oil Corporation (CNOOC), which was already founded in 1982 and, since then, has been concentrating on offshore investments, these three state owned companies form today's core of the Chinese oil industry.[4]

Though the Chinese government is the majority shareholder in all three companies, the recent years have shown that CNPC, Sinopec, and CNOOC are not mere puppets in the hands of the political elites. On the one hand, large investments abroad of the national oil companies are often preceded with frequent visits of Chinese diplomats, who usually carry along large funds for infrastructure projects or new presidential palaces. On the other hand, competition between the three companies has tightened in spite of the discomfort of the central government. In 2005, for instance, Sinopec and CNPC bid for the same pipeline project in Sudan and, hence, competed directly against each other. Due to surging profits, their reliance on international banks, and the listings of several subsidiaries on foreign stock exchanges, the national oil companies have gained considerable power vis-à-vis the central government and

---

[2] Aside from oil, China is mainly relying on coal-fired energy production (it has the world's third largest reserves), but very little on natural gas. Cf. BP p.l.c., 2010 pp. 27-29; Lee, et al., 2008 pp. 114-115
[3] Zweig, et al., 2005 p. 36
[4] Lee, et al., 2008 pp. 113-114; Sieren, 2008 p. 16

its agencies. Nevertheless, one should not expect that any of the national oil companies would act directly against the will of the political leadership.[5]

2.2.   Statistical Evidence for China's Foreign Oil Needs

> *"Developing nations, including population giants China and India, are entering their most energy-intensive phase of economic growth as they industrialise, build infrastructure, and increase their use of transportation."*[6]

25 years ago, the People's Republic of China was East Asia's largest oil exporter. This has changed dramatically. China has shifted from being an oil exporter to being the world's second largest importer of oil.[7] The figures of the table below underline this fact. The proven reserves of oil in China have declined slightly in the last decade while the total consumption increased by over 80 percent. Today, more than ten percent of the oil consumed in the world is consumed in China. Considering that the per capita oil consumption in China still lacks far behind the consumption in the western hemisphere and that China makes up about 20 percent of the world population, this ratio is likely to increase beyond 10.4 percent.[8]

China	1990	2000	2001	2002	2003	2004	2005	2006	2007	2008	2009	Share of total
Proven reserves (in bill. barrels)	16.0	15.1									14.8	1.1%
Total consumption (in '000 barrels daily)		4772	4872	5288	5803	6772	6984	7410	7771	8086	8625	10.4%

Table: China's Oil Sector[9]

Foreign resources in general and oil in specific are necessary to fuel China's ongoing economic growth, which has averaged on an annual increase in GDP of nine percent over the last three decades. This continuing growth is not just important to lift more and more people

---

[5] Rotberg, 2008 pp. 4-5; Downs, 2007 pp. 48-51
[6] Shell International BV, 2008 p. 8
[7] Zweig, et al., 2005 p. 25
[8] BP p.l.c., 2010 pp. 6-15
[9] BP p.l.c., 2010 pp. 6, 12

out of poverty, but also to prevent social unrest and, therefore, to solidify the sole leadership of the Chinese Communist Party. The African continent has become an important partner for the People's Republic and the Communist Party to achieve these goals.[10]

2.3.   Trends in the African Oil Industry

The distribution of the world wide proven oil reserves is, not surprisingly, in favor of the Middle East. More than half of the total reserves are situated in that region. Nevertheless, Africa also has considerable stakes in the game. In 1990, 5.9 percent of the proven oil reserves were located in Africa; this percentage has risen to 9.6 percent in 2009. The growth in this number indicates why Africa might become even more attractive for international oil companies as it is today. Africa has the largest potential in discovering undetected oil reserves; it is one of the few regions in the world where large oil fields may still be hidden.[11]

Africa	1990	2000	2001	2002	2003	2004	2005	2006	2007	2008	2009	Share of total
Proven reserves Africa (in bill. barrels)	59.1	84.7									127.5	9.6
Proven reserves Sudan (in bill. barrels)	0.3	0.3									6.7	0.5%
Oil production Sudan (in '000 barrels daily)		174	217	241	265	301	305	331	468	480	490	0.6%

Table 1: Africa's Oil Sector[12]

Within Africa, Libya and Nigeria have the largest reserves on hand and make up about two thirds of Africa's total reserves; third and fourth place take Algeria and Angola with roughly

---

[10] Friedberg, 2006 p. 34; Zweig, et al., 2005 pp. 25-26

[11] Between 2001 and 2004, 85 percent of the world's newly found oil reserves were located in West and Central Africa. The proven reserves in Africa increased by 56 percent between 1996 and 2006. In comparison, in the rest of the world the reserves increased by only 12 percent in the same period of time. Raine, 2009 p. 38; Downs, 2007 p. 45; BP p.l.c., 2010 pp. 6-7

[12] BP p.l.c., 2010 pp. 6, 8

5

ten percent each. Sudan is on fifth place with a share of approximately five percent. However, the country can present a strongly growing oil production throughout the last decade.[13]

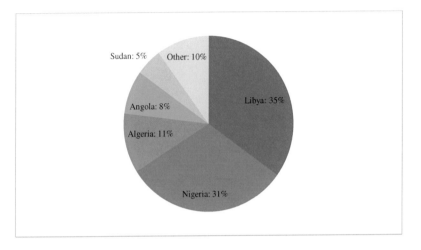

Figure 1: Proven Oil Reserves in Africa[14]

With the exception of Sudan, where the oil sector is dominated by Chinese national oil companies, especially by CNPC, China remains a relative minor player in the African oil market. Among foreign oil companies, the commercial value of oil investments of China's oil companies amount to only eight percent; if we include African companies in the equation it amounts to only three percent.[15]

---

[13] Downs, 2007 pp. 43-45
[14] Downs, 2007 p. 43
[15] Trinh, et al., 2006; Downs, 2007 p. 44

3. China & Sudan: A Partnership of Convenience

3.1. A Historical Overview of China's Ties to Africa and the Sudan

For almost 60 years, the People's Republic of China has systematically cultivated solidarity and a well aligned set of diplomatic as well as economic relations to many African nations. Initially, the People's Republic wanted to win new diplomatic recognition to take over the permanent seat of the Taiwan based Republic of China in the UN Security Council. As more and more African nations won their independence, the People's Republic saw its chance to gain allies to support its bid at the annual General Assembly. Finally, at the 26th UN General Assembly in 1971 this policy was met with success. 26 African states supported the People's Republic motion recognizing it to be the only lawful representative of China to the UN. [16] Today, of course, China cultivates it ties to Africa for a variety of reasons: *"China needs Africa. It needs Africa for resources to fuel China's development goals, for markets to sustain its growing economy and for political alliance to support its aspirations to be a global influence."*[17]

China's predominantly diplomatic relations with the Sudan started to shift to a much more economically based relationship, when the China National Petroleum Corporation (CNPC) replaced a Canadian enterprise within an international consortium, that had been granted oil exploration rights from the Sudanese government. Initially, the first oil company that started exploration in Sudan was the US based giant Chevron in 1974. After continuous tensions, civil war between the Christian south and the more populous Islamic north broke out in 1983. Hence, the security situation deteriorated so severely that Chevron decided to sell its interests to the international consortium GNPOC, which was dominated by a Canadian firm.[18] Though, there had been a ceasefire since 1995, the security situation remained tense. Moreover, when Omar al-Bashir threw over the former leadership through a military coup and established an Islamic government based on Sharia law in 1989, the world's public opinion turned against the Sudanese leadership and western companies that potentially backed its reign through petro-dollars. The Chinese CNPC finally saw its chance to step in and take over the stakes of the Canadian firm in the GNPOC-consortium in 1997.[19]

By the end of the 1990s the Sudan was in dire straits. Its debt was sky rocking and the ongoing civil war had wracked the local economy. To increase its oil revenues was a logical

---

[16] Raine, 2009 pp. 14-19
[17] Gill, et al., 2007 p. 9
[18] GNPOC stands for Greater Nile Petroleum Operating Company
[19] Engdahl, 2007; Lee, et al., 2008 pp. 124-126

7

step and maybe the only escape from this desperate financial situation. Since all multinational oil companies had left the country earlier, CNPC was the only company in a position to fill this vacuum.[20] Therefore, the Sudanese government agreed to sell large oil concessions to CNPC on an equity basis. Though oil was at first discovered in Sudan in 1974 by Chevron, 25 years later in 1999 the country exported its first barrel of oil thanks to the Chinese investment. Since then, oil production has steadily increased from 174,000 barrels a day in 2000 to 490,000 barrels a day in 2009.[21] In spite of political and social instability, Sudan offers one major advantage that attracts Chinese oil companies: Sudan allows foreign companies equity access in a world in which more than three quarters of the oil reserves are closed to such investments.[22]

3.2. China's Modus Operandi in Securing Sudanese Oil

China has been pursuing somewhat of a scheme in securing oil reserves on the African continent and in other parts of the world.[23] Usually, Chinese oil companies bring with them a whole entourage of other Chinese enterprises including construction companies, trading firms and the Eximbank,[24] which facilitates import and export activities and grants concessional loans.[25] Since Chinese oil companies cannot offer cutting edge technologies like the multinational oil corporations, they often offer package deals to resource rich developing countries. These package deals may contain the financing and conduct of infrastructure projects, investments in other economic sectors of the host country, training to local employees, and low interest loans. On top of that, the Chinese government usually cultivates diplomatic strings to the host country, including the granting of development aid.[26]

Chinese non-oil investments in Sudan are significant and include a large hydroelectric facility, a new airport in Khartoum, and several textile plants. Over 25,000 Chinese workers are living in Sudan today. In fact, more than half of Sudan's total exports go to China. Furthermore, China operates almost all of Sudan's oil production and has a 50 percent share in the

---

[20] Zweig, et al., 2005 p. 32
[21] Lee, et al., 2008 pp. 124-126; for data about oil production in Sudan see also Table 2 above.
[22] Downs, 2007 p. 45; Lee, et al., 2008 pp. 112, 132
[23] For a comprehensive overview of China's worldwide activities to secure resources and precious metals cf. Friedberg, 2006.
[24] By 2008, the Eximbank was financing approximately 300 projects in 36 African countries; cf. Raine, 2009 pp. 27-28
[25] Concerning the Chinese donor system, the thresholds between common credit financing, concessional loans, and development aid tends to be rather elusive. For a detailed approach to differentiate Chinese lending practices cf. Brautigam, 2008 and Hubbard, 2008.
[26] Rotberg, 2008 pp. 1-3; Zweig, et al., 2005 pp. 27-30

country's only major oil refinery. Since western companies have left the country, China faces hardly any competition in securing Sudan's oil riches. Moreover, China has the perception that almost all other oil producing regions are already tied to US, European, or Japanese interests, and thus, might be less receptive to Chinese offers.[27]

3.3.   China's Justification for its Sudan Policy

In the past, China has been criticized by many nations, which tried to marginalize Sudan due to its government's atrocities and human rights violations. Nevertheless, China has expanded its economic activities in the Sudan over the last 13 years, regularly blocked resolutions against the Sudanese leadership at UN level, and did not join in public critique against the regime. [28] China justifies its ongoing close relation to Sudan and its ruling elite with a broad set of arguments:

First of all, China emphasizes that through the various investments the PRC makes in the Sudan, it provides a major cornerstone to develop the country and its economy. China does not solely build roads and other means of infrastructure but also grants favorable loans for national priorities, trains local people, and thus stimulates the domestic economy. Economic development is perceived to be the only effective remedy to resolve the poverty-rooted Darfur crisis. Critics of Chinese investment practices in Sudan, on the other hand, argue that Chinese investments are usually performed by Chinese enterprises, which mainly employ Chinese people, and that these investments are mostly made to pursue Chinese interests. Furthermore, they worry that granting extensive loans to Sudan, on whatever favorable conditions, may end up in a debt entanglement such as many African countries have experienced in the past.[29]

Secondly, the PRC likes to portray itself as a role model for Africa. China freed itself from colonial bondages to survive internal chaos and economic misery. In the aftermath the country achieved spectacular economic growth and lifted millions of people out of poverty. There are some undeniable analogies to the African continent and the Sudan. This Chinese success story finds a large resonance in Africa.[30]

Lastly, when being criticized for its Sudan policy, China always refers to its often stated principle of non-interference. In 2004, then-deputy foreign minister Zhou Wenzhong said in

---

[27] Lee, et al., 2008 pp. 124-126
[28] Zweig, et al., 2005 pp. 32-33; Downs, 2007 pp. 58-59
[29] Downs, 2007 pp. 60-61; Zweig, et al., 2005 p. 32; Lee, et al., 2008 pp. 116-117
[30] Gill, et al., 2007 pp. 7-8

an New York Times interview, when asked about Sudan's human rights record, that, *"We try to separate politics from business. Secondly, I think the internal situation in the Sudan is an internal affair, and we are not in a position to impose upon them."*[31] The Chinese principle of non-interference, in fact, dates back to 1954 when the Communist Party first formulated five key principles for the People's Republic foreign affairs policy. The 'Five Principles of Peaceful Coexistence' are still valid today in China's official political philosophy and include aside from non-interference, mutual respect for sovereignty, non-aggression, equality, and peaceful coexistence. Ever since, the Chinese leadership uses the principle of non-interference to shield criticism against domestic human rights abuses, business relations to rogue regimes, and presumably internal issues, as for instance the Taiwan-dispute.[32]

---

[31] French, 2004
[32] Raine, 2009 pp. 15-16; Gill, et al., 2007 pp. 12-13

4. China's Future Oil Strategy in Sudan

4.1. Threats for China's Oil Strategy

China currently faces several threats that could adversely affect its oil strategy in Sudan. Smoldering conflicts between the northern part of Sudan and the south entail considerable risks for Chinese workers, the country's oil infrastructure, and thus, for China's investment strategy. Since the central government has financed the civil war and its weapon arsenal mainly with oil money, cutting Khartoum off the oil source is an obvious approach to constrain its power. Indeed, pipelines to Sudan's only refinery near the capital have been attacked frequently in the past.[33] Although a peace agreement was signed from both sides in 2005, the situation remains tense, especially as the secession vote, which is scheduled for January 2011, approaches. After a six-year period of autonomous self-rule, the south shall decide through a public vote whether to break away from Sudan or not. In this context new clashes between the two opponent parties may occur. A new civil war would bring with it a deterioration of the security situation and would exacerbate access to the oil fields, which are located in the southern part of the country.[34]

The Darfur conflict is another matter that could imply high strategic costs for China. Though, as the map below indicates, there are no noteworthy oil reserves within the boundaries of Darfur, the humanitarian crisis in the region has aroused the world's attention. Since 2003, Khartoum has been fighting against African tribes that feel neglected by the central government and revolt against it. Furthermore, regime-backed Arabian militias[35] roam on horsebacks through the region to murder and displace Black Africans. Many believe that China fuels this conflict because it not only backs the Sudanese regime and is the largest buyer of its oil, but also sells weaponry to it. In fact, it turned out that most small arms weapons used by the militias were of Chinese made. The Darfur crisis has harmed China's international reputation, and hence, could impair its political clout in the world.[36]

---

[33] Lee, et al., 2008 pp. 126-128

[34] Engdahl, 2007; Rotberg, 2008 pp. 12-13

[35] The infamous militias are called Janjaweed and consist mainly of members from different Arabian tribes from the northern part of Darfur. Most estimates assume that the conflict so far accounts for a death toll of approximately 200,000 people, with some 2,000,000 displaced. Cf. Engdahl, 2007; Brown, et al., 2008 pp. 256-260; Mükke, 2010

[36] Brown, et al., 2008 pp. 256-260

4.2. The Two States Scenario

If the secession vote indeed is going to take place next year, the probability that the south will break away from Sudan is rather high. One decisive question is where a possible future border would separate the then two states.[37]

The largest part of the oil fields is located in the south. So in a two states scenario Khartoum could no longer dispose all alone over the whole oil reserves. Suddenly, there would be another authority that was to decide over at least part of the oil. Chinese diplomats have once more proved to be quick in adjusting to new situations. They have already built up relations to the autonomous government in Juba. In fact, the Chinese are among the main investors to rebuild the capital of southern Sudan, which has been devastated by decades of war. Though, it is still in most people's memory that China backed the regime in Khartoum through petro-dollars, and thus, helped it to finance the war against the south, Juba, just like many other governments of resource rich developing countries, appreciates the quick and uncomplicated development aid and investments from the People's Republic of China.[38]

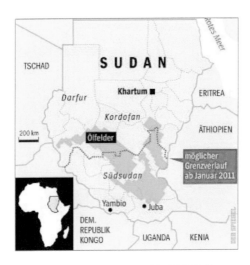

Figure 2: Map of Sudan and its Oil Fields[39]

[37] Mükke, 2010
[38] Lee, et al., 2008 pp. 130-133
[39] The map is extracted from a Spiegel article: Mükke, 2010

China and South Sudan, though, face one major problem concerning the oil infrastructure. Until now, the only way to export oil from the Sudan to China or any other part in the world is via the port of Masra al Bashair in the north. If the south secedes from the north, the new southern state will be landlocked with no access to the Red Sea. Presumably it would be politically difficult, if not impossible, for the south to further ship its oil through Masra al Bashair. In that context the Chinese are even more attractive for Juba, because they are working with the Kenya Pipeline Corporation to explore the possibility to build a pipeline from the oil fields located in southern Sudan to the port of Lamu in Kenya. Such a pipeline would grant South Sudan a strong bargaining power vis-à-vis Khartoum, and would make it less dependent from the abhorred regime. First estimates ticket the costs to about $ 1.4 billion. Indeed, such a gigantic project could be only shouldered with the help of the Chinese.[40]

---

[40] Lee, et al., 2008 pp. 130-133; Mükke, 2010

5. Conclusion

China's entry into Africa and the Sudan has been a team effort. The national oil companies, the country's export-import bank, economic and trade agencies, and key trading enterprises all play an important role in China's resource and oil strategy in the East African country.

Sudan is among the People's Republic's most important providers of oil. In the mid-1990s, China took the chance to slip in and replace the multinational oil giants in the Sudan. They left the country together with nearly all other western companies because of the regime's autocratic governance and severe human rights violations associated with it. Hence, there are two decisive reasons why China went for such a large scale investment in the Sudan. First of all, the PRC faces almost no competition from the multinationals, which are in terms of technology still far ahead of the Chinese national oil companies. Secondly, China wants to hedge economic risks with secure oil supplies. It might be politically risky to go after Sudan's oil, but since Khartoum allows China large equity stakes in the whole oil-related value chain including in the oil fields itself, economically speaking, Sudan's oil is almost as valuable as domestic reserves.

However, China had to pay a strategic price in securing Sudan's oil. The PRC has been strongly criticized for tolerating Khartoum's despotic rule and therefore has suffered from a considerable loss of reputation. Nonetheless, in the run-up to the Olympic Games in Beijing, one could observe a slight shift in China's Sudan policy. So far, China had been always insisting on its principle of non-interference and even blocked UN resolutions against the regime. In 2007, however, China persuaded Sudan's ruling elite to allow peacekeeping troops from the UN and the African Union in Darfur. Of course, China did not publicly criticize Khartoum but used a more subtle approach behind closed doors. That Sudan's regime caved in and accepted such troops on its soil shows that China has leverage over Khartoum other nations, most notably the US, do not have. The future will show if China uses this leverage to push Khartoum to a less ruthless governance or if the People's Republic will lag behind its own ambitions to become the leading actor on the international stage.

BIBLIOGRAPHY

BP p.l.c. 2010. BP Statistical Review of World Energy - June 2010. *BP Statistical Review.* [Online] June 2010. [Cited: September 12, 2010.] www.bp.com/statisticalreview.

Brautigam, Deborah. 2008. China's Foreign Aid in Africa: What Do We Know? [ed.] Robert I. Rotberg. *China into Africa: Trade, Aid, and Influence.* Washington D.C. : Brookings Institution Press, 2008, pp. 197-216.

Brown, Stephen and Lekha Sriram, Chandra. 2008. China's Role in Human Rights Abuses in Africa: Clarifying Issues of Culpability. [ed.] Robert I. Rotberg. *China into Africa: Trade, Aid, and Influence.* Washington D.C. : Brookings Institution Press, 2008, pp. 250-271.

Downs, Erica S. 2007. The Fact and Fiction of Sino-African Energy Relations. *China Security.* 2007, Vol. 3, 3, pp. 42-68.

Engdahl, William F. 2007. Darfur? It's the Oil, Stupid...: China and USA in New Cold War over Africa's oil riches. *Geopolitics - Geoeconomics-website.* [Online] May 20, 2007. [Cited: September 7, 2010.] www.engdahl.oilgeopolitics.net/print/China%20&%20US%20in%20Cold%20War%20over%20Afric a's%20Oil.html.

French, Howard. 2004. China in Africa: All Trade, With No Political Baggage. *The New York Times.* August 8, 2004, p. 4.

Friedberg, Aaron L. 2006. "Going Out": China's Pursuit of Natural Resources and Implications for the PRC's Grand Strategy. *NBR Analysis.* September 2006, Vol. 17, 3.

Gill, Bates, Huang, Chin-hao and Morrison, J. Stephen. 2007. Assessing China's Growing Influence in Africa. *China Security.* 2007, Vol. 3, 3, pp. 3-21.

Hanson, Stephanie. 2008. China, Africa, and Oil. *Council on Foreign Relations.* [Online] June 6, 2008. [Cited: September 7, 2010.] www.cfr.org/publication/9557/china_africa_and_oil.html.

Hubbard, Paul. 2008. Chinese Concessional Loans. [ed.] Robert I. Rotberg. *China into Africa: Trade, Aid, and Influence.* Washington D.C. : Brookings Institution Press, 2008, pp. 217-229.

Lee, Henry and Shalmon, Dan. 2008. Searching for Oil: China's Oil Strategies in Africa. [ed.] Robert I. Rotberg. *China into Africa: Trade, Aid, and Influence.* Washington D.C. : Brookings Institution Press, 2008, pp. 109-136.

Mükke, Lutz. 2010. Nord gegen Süd: Warum der Sudan zerbricht. *Spiegel Online.* [Online] April 15, 2010. [Cited: September 24, 2010.] http://www.spiegel.de/politik/ausland/0,1518,687830,00.html.

Raine, Sarah. 2009. China's African Challenges. *Adelphi Papers.* 2009, Vol. 49, 404, pp. 13-142.

Rotberg, Robert I. 2008. China's Quest for Resources, Opportunities, and Influence in Africa. [ed.] Robert I Rotberg. *China into Africa: Trade, Aid, and Influence.* Washington D.C. : Brookings Institution Press, 2008, pp. 1-20.

Saam, Wolfgang. 2008. Chinas Griff nach Afrikas Rohstoffen: Auswirkungen auf Afrikas Entwicklung und Europas Versorgungssicherheit. *Analysen und Argumente.* January 2008, 49.

Schüller, Margot and Asche, Helmut. 2007. China als neue Kolonialmacht in Afrika? Umstrittene Strategien der Ressourcensicherung. *GIGA Focus.* [Online] February 2007. [Cited: September 10, 2010.] www.giga-hamburg.de/giga-focus. ISSN 1862-3581.

Shell International BV. 2008. *Shell Energy Scenarios to 2050.* The Hague : s.n., 2008.

Sieren, Frank. 2008. Der Coup. *Die Zeit.* October 16, 2008, 43, p. 16.

Trinh, Tamara, Voss, Silja and Dyck, Steffen. 2006. Chinas Rohstoffhunger: Auswirkungen auf Afrika und Lateinamerika. *Deutsche Bank Research.* [Online] June 30, 2006. [Cited: September 7, 2010.] www.dbresearch.de. ISSN Internet: 1435-0734.

Zweig, David and Jianhai, Bi. 2005. China's Global Hunt for Energy. *Foreign Affairs.* September 2005, Vol. 84, 5, pp. 25-38.

Lightning Source UK Ltd.
Milton Keynes UK
UKRC02n1024151117
312757UK00001B/1

* 9 7 8 3 6 5 6 0 5 0 5 9 9 *